EYE ON SPACE

Space Watch: Planets

Chris Oxlade

PowerKiDS press.
New York

Published in 2011 by The Rosen Publishing Group Inc.
29 East 21st Street, New York, NY 10010

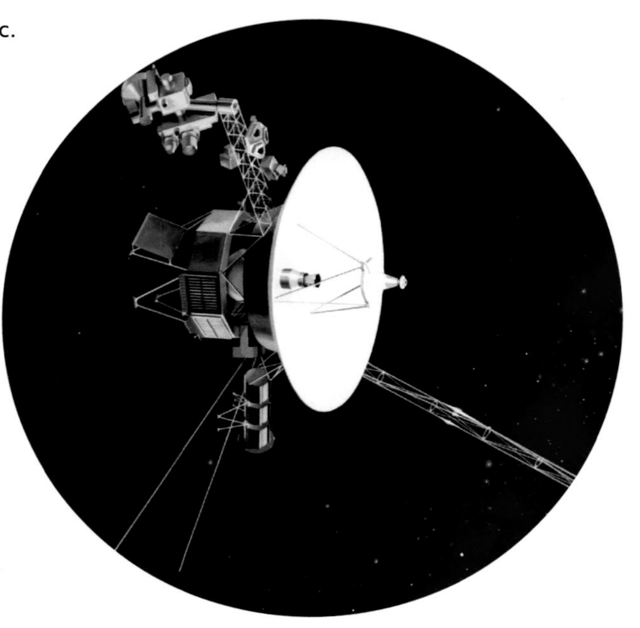

First Edition

Editor: Julia Adams
Designer: Robert Walster
Picture researcher: Julia Adams

Library of Congress Cataloging-in-Publication Data

Oxlade, Chris.
Space watch. Planets / by Chris Oxlade. — 1st ed.
p. cm. — (Eye on space)
Includes index.
ISBN 978-1-61532-542-9 (library binding)
ISBN 978-1-61532-548-1 (paperback)
ISBN 978-1-61532-549-8 (6-pack)
1. Planets—Juvenile literature. I. Title. II. Title: Planets.
QB602.O95 2011
523.4—dc22

2009044620

Photographs:
Andy Crawford: 22, 23; Alamy: Martin Harvey 4, Suzanne Long 10, Mark Garlick 18; Lunar and Planetary Laboratory: OFC; NASA/JPL: 2, 19; NASA/Goddard Space Flight Centre: 5; Shutterstock: oorka 14; Science Photo Library: David A. Hardy/Futures 50 Years in Space 11, JPL/NASA 20, Roger Harris 1, 21, The International Astronomical Union/Martin Kornmesser 6/7, NASA/Johns Hopkins University Applied Physics Laboratory/Carnegie Institution of Washington 8, Mark Garlik 9, NASA 13, 15, 16, NASA/ESA/STSCI/ Hubble Heritage Team 12.

Manufactured in China
CPSIA Compliance Information: Batch #WAS0102PK: For Further Information contact Rosen Publishing, New York, New York at 1-800-237-9932

Web Sites

Due to the changing nature of Internet links, PowerKids Press has developed an online list of Web sites related to the subject of this book. This site is updated regularly. Please use this link to access this list:
http://www.powerkidslinks.com/eos/planets

Contents

Planet Earth

The Earth is the place where we live.
Its surface is made up of land and water.

Most of the Earth's surface is covered with seas and oceans.

The Earth is a planet. From space, it looks like a giant ball. The shape of the Earth is called a sphere.

The Earth is sometimes called the Blue Planet. Why do you think this is?

The Solar System

The Earth is part of a group of eight planets. All the planets in this group travel around the Sun.

Earth is the third planet from the Sun.

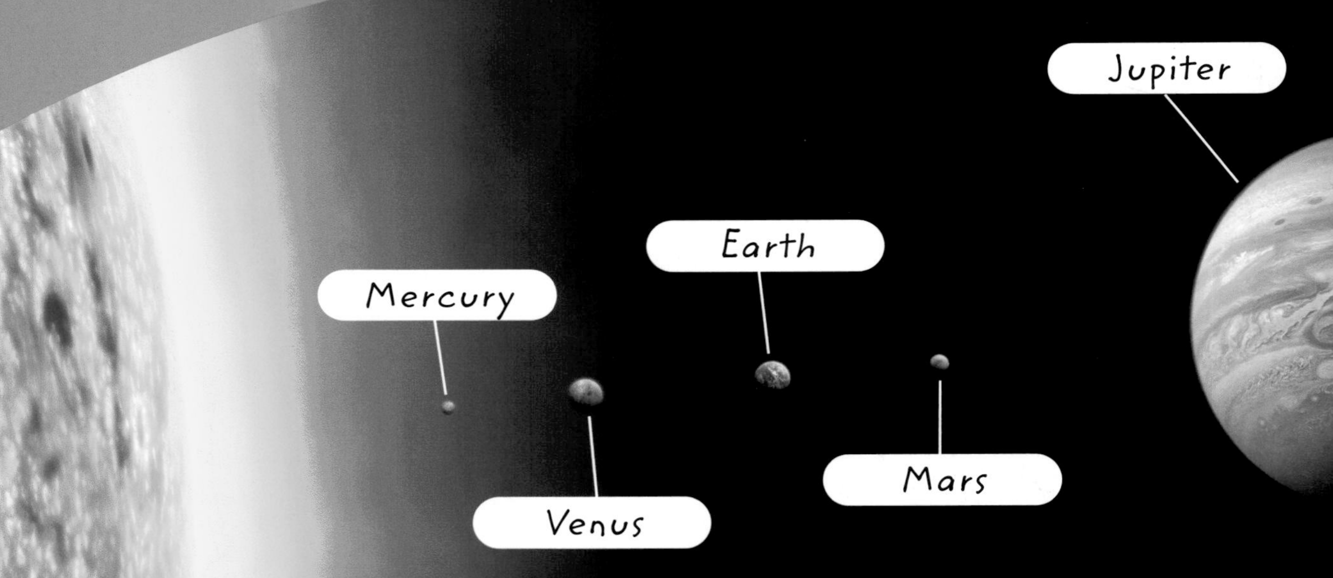

The planets are each at a different distance from the Sun. Together, the Sun and the planets are called the solar system.

Mercury

The closest planet to the Sun is called Mercury. It is also the smallest planet in the solar system.

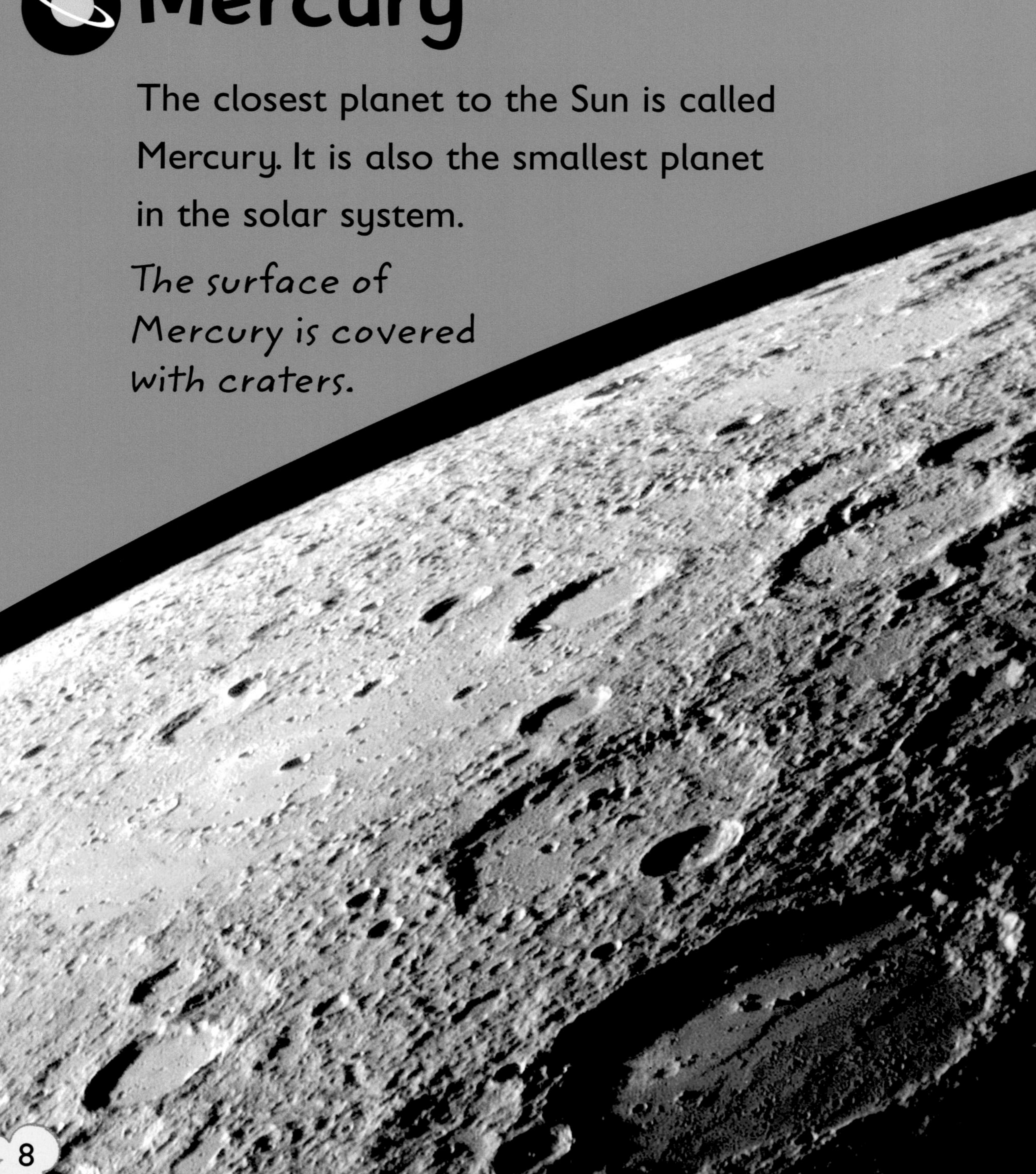

The surface of Mercury is covered with craters.

The side of Mercury that faces the Sun is extremely hot. The other side is freezing cold.

When the Sun shines on Mercury, it is hot enough on the surface to melt metal!

Venus

The closest planet to Earth is called Venus. You can sometimes see Venus in the morning or evening sky.

Venus looks like a very bright star in the sky.

Venus is covered with thick clouds.
The clouds trap heat from the Sun.
This makes Venus the hottest planet
in the solar system.

The sky on Venus is yellow and cloudy.

Mars

The surface of Mars is covered with red rocks and dust. This is why Mars is also called the Red Planet.

The top and bottom of Mars are covered with ice.

Many spacecraft have visited Mars. Some have landed on the surface to explore.

The Mars Exploration Rover landed on Mars in January 2004.

The rocks on Mars are red because they are rusty. Do you know what rust is?

Jupiter

The solar system's largest planet is Jupiter. It is so big that all the other planets could fit inside it.

Jupiter makes the Earth look tiny.

Jupiter is made of gas and liquid.
It is covered with bands of cloud.
There are giant storms in the clouds.

This swirl of cloud is called the Great Red Spot.

Saturn

The giant planet Saturn is made of gas and liquid. Its surface is covered with bands of clouds.

This red mark is a thunderstorm on Saturn called the Red Dragon.

Saturn is famous for its rings. The rings are made up of millions of lumps of ice.

The black lines are the gaps between the rings.

Uranus

The surface of Uranus is very cold, because it is so far away from the Sun.

Uranus has some rings, but most of them are too dark to see.

The spacecraft Voyager 2 visited Uranus in the year 1986. It took nine years to travel from Earth to Uranus.

Voyager took pictures of Uranus and sent them back to Earth.

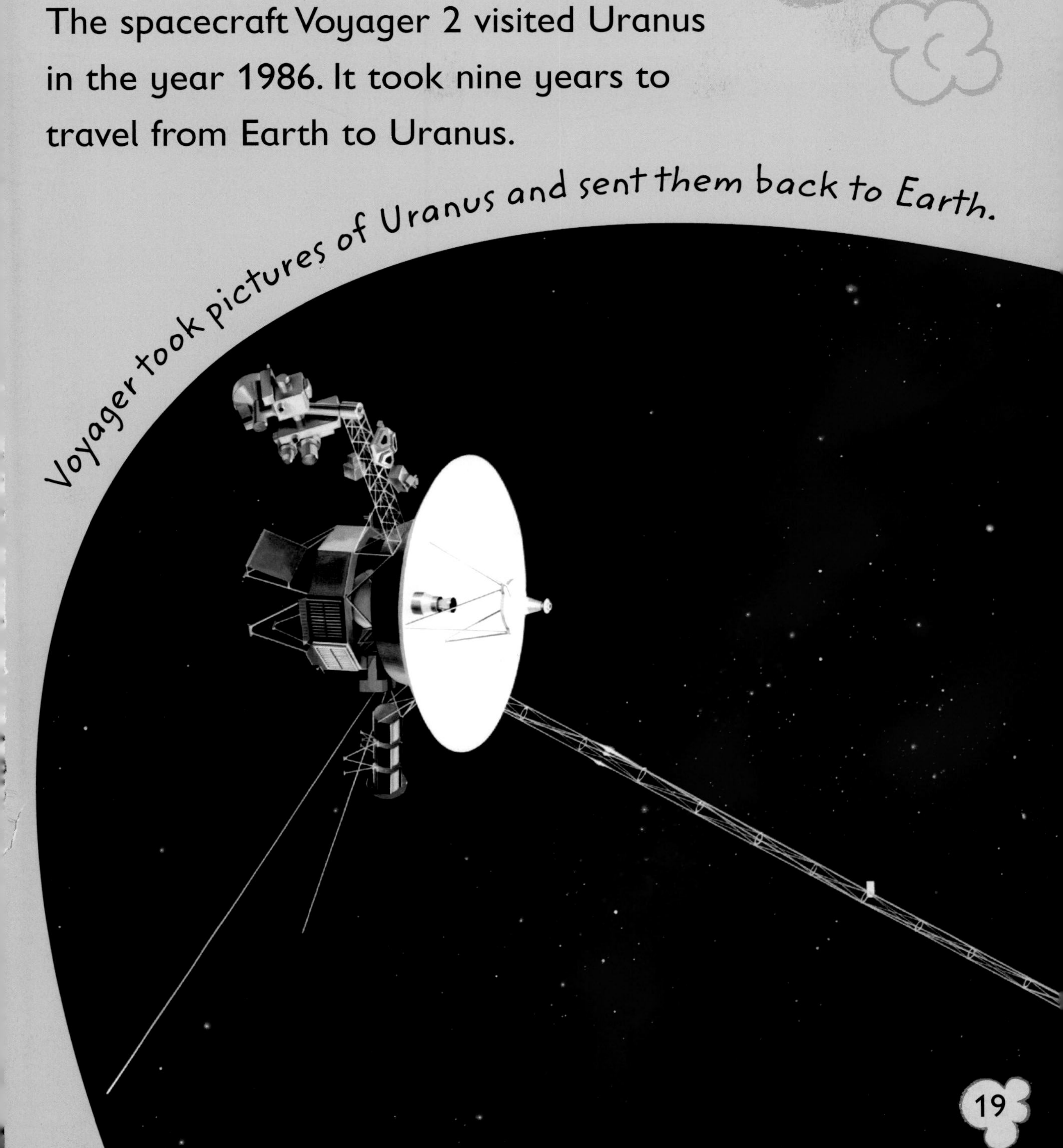

Neptune

Neptune is the coldest planet in the solar system. It is about the same size as Uranus. It has some swirling clouds and storms.

From space, the storms on Neptune look like white streaks.

Neptune is the farthest planet from the Sun. It is named after the Roman god of the sea, because it looks blue.

From Neptune, the Sun looks like a dot of light.

Neptune is the stormiest planet in the solar system.

Planets Mobile

Make a mobile of the eight planets to hang up in your room!

You will need:

- white card
- ruler
- scissors
- tape
- paints
- string
- hole punch

Mercury:	¾ inch (2 cm)	Jupiter:	11 inches (28 cm)
Venus:	2 inches (5 cm)	Saturn:	9½ inches (24 cm)
Earth:	2 inches (5 cm)	Uranus:	4 inches (10 cm)
Mars:	1¼ inch (3 cm)	Neptune:	4 inches (10 cm)

1. Draw eight circles. These will be the planets Check the list above to find out how wide they need to be. You can use different round objects to trace around.

2. Cut out each circle. Write the name of each planet on the back with a pencil.

3. Paint each planet. Look at the photographs in this book to find out the right colors and patterns for each of them.

4. Use the hole punch to make one hole in each planet.

5. Use a long piece of string to thread all your planets. Make sure they are in the right order.

6. Now you can hang your solar system up in your room!

To add the Sun to your mobile, you will need to add a circle that is 5¾ feet (1.75 m) wide.

Glossary and Further Information

crater a dish-shaped hole in the surface of a planet

gas a gas has no shape and can be invisible. Air is a mixture of gases

liquid a runny material that takes the shape of its container

metal a strong, hard material that is often shiny

planet a giant ball of rock or liquid in space

rust a red substance. Rust forms when the metal iron is left in water for a long time

spacecraft a machine that travels through space

Books

Boy, Were We Wrong About The Solar System
by Kathleen V. Kudlinsk
(Dutton Juvenile, 2008)

Living in Space
by Alex Pang
(Usborne, 2006)

Our Solar System
by Seymour Simon
(Collins, 2007)

Index